Robotic Fish iSplash-OPTIMIZE:
Optimized Linear Carangiform Swimming Motion

By
Richard James Clapham PhD

iSplash Robotics

Published by *iSplash* Robotics
www.isplash-robotics.com
r.j.c@ieee.org

Copyright © 2016
Published by *iSplash* Robotics 2016
Illustration by Richard James Clapham PhD

All rights reserved. No part of this publication may be reproduced, stored in a retrieval system or transmitted in any form or by any means, electronic, mechanical, photocopying or otherwise without the prior permission of *iSplash* Robotics.

ISBN-13: 978-1537269993
ISBN-10: 1537269992

Thank you for your purchase.

Abstract—This paper presents a new robotic fish, *iSplash*-OPTIMIZE, which is 0.6m in body length and deploys a single actuator to drive discrete links across the full-body length. The main focus is on optimizing the kinematics parameters of its linear carangiform swimming motion in order to improve the distance travelled per beat. The experimental results show that the fish can be actuated at high frequencies up to 20 Hz due to deploying a continuous rotary power source. Each discrete link is able to be precisely tuned, providing accurate kinematics with little mechanical loss.

Keywords: Robotic fish · Carangiform swimming · Mechanical drive system · Full-Body length.

1 Introduction

Live fish are able to generate locomotive forces resourcefully in comparison to underwater vehicles (UVs) powered by rotary propellers [1,2,3,4]. The fish locomotion is able to extract energy from upstream vortices. A passive body resonating within the *Karman Vortex Street* can generate forward locomotion, with the highest locomotive efficiency [5].

A prominent parameter for analyzing body and caudal fin locomotive performance is the distance traveled per body length during one caudal fin oscillation. This can be calculated using the Swimming number (*Sw*) defined as *Sw=U/fl* where U denotes swimming velocity, f denotes the tail beat frequency and l denotes the body length [6]. The distance travelled per tail beat of the Cyprinus carpio (common carp) is highly efficient measuring a *Sw* of 0.66, with the highest recorded example of the Tursiops (Bottlenose dolphin) with a *Sw* of 0.82.

Research into biomimetic underwater propulsion has developed several innovative hydrodynamic mechanisms, such as Barrett's Robotuna which achieved a maximum velocity of 0.65 body lengths/ second (BL/s) (i.e. 0.7m/s) [7], Yu's prototype achieving a maximum velocity of 0.8BL/s (i.e. 0.32m/s) [8], Essex's G9, achieving a maximum velocity of 1.02BL/s (i.e. 0.5m/s) [9] and Valdivia y Alvarado's build achieving a maximum velocity of 1.1BL/s (i.e. 0.32m/s) [10]. To date, the speed of robotic fish is approximately 1BL/s, but real fish have an average maximum velocity of 10BL/s with the common carp peaking at 12.6 BL/s [3], [11].

Our first prototype *iSplash*-I [12], which is 0.25m in body length, weighs 0.345kg, and has a tethered power supply, achieved a high performance carangiform swimming motion. Its novel mechanical drive system operated in two swimming patterns, a traditional posterior confined undulatory swimming pattern and the introduced coordinated full-body length swimming pattern. The proposed swimming motion significantly improved the accuracy of the kinematic displacements and greatly outperformed the posterior confined approach in terms of speed, achieving 3.4BL/s and consistently achieving a maximum velocity of 2.8BL/s at 6.6Hz with a low energy consumption of 7.68W. Based on intensive observation, the predetermined offsets of the discrete link assembly were set to generate the kinematic parameters of a common carp.

Optimizing the discrete structure was limited as only the final link of over the full-body length was able to be adjusted. Experimental testing found that its Sw of 0.42 was lower than a common carp's. The coordinated full-body length swimming motion provided accurate kinematics directly related to increased swimming performance, we estimate that further tuning of the kinematic parameters may provide a greater increase in maximum velocity and reduce the cost of transport. In consideration of this *iSplash*-OPTIMIZE was proposed, shown in Figs. 1 and 2.

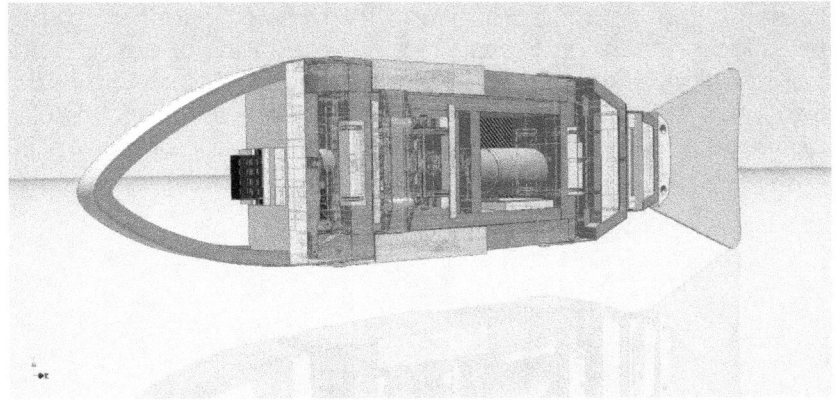

Fig. 1. *iSplash*-OPTIMIZE

This research aims to improve the efficiency of the linear swimming motion with eight main objectives: (i) to develop a new build mechanical drive system, capable of distributing power from a single actuator to discrete links across the full-body length; (ii) to allow each discrete link of the assembly to be precisely tuned by devising a powertrain with innumerable adjustments; (iii) to attain a structurally robust mechanical drive system capable of intensively high frequencies of 20Hz;

(iv) to devise a prototype able to carry a high capacity power supply to obtain consistent high force free swimming; (v) to gain high speeds by developing a compact design taking into consideration geometric and kinematic parameters throughout the full tail beat cycle; (vi) to obtain a low cost of transport by developing a mechanical drive system considering methods that may cause internal mechanical losses; (vii) to deploy an electrical system accurately measuring energy consumption, destabilization, with perception sensing, wireless communication and ability to produce control signals for multiple actuators for future autonomy; (viii) to validate the study by conducting a series of experiments measuring the prototype's achievements in terms of energy consumption and kinematic parameters.

The remainder of the paper is organized as follows: Section **2** presents the linear carangiform swimming patterns to be investigated. Section **3** describes the mechanical design and construction method of the new build *iSplash*-OPTIMIZE. Section **4** describes the experimental procedure and results obtained. Concluding remarks and future work are given in Section **5**.

2 Linear Swimming Patterns to be Investigated

2.1 The Traditional Swimming Motion Approach

This study aims to significantly improve the accuracy of replicating the wave form of the carangiform swimming mode, specifically the common carp, due to its high locomotive performance [6]. Identified by the portion of the body length actively displaced, the selected carangiform applies the swimming method of body and/or caudal fin propulsion, associated with the method of added mass. A full description can be found in [13]. Fish have an improved cost of transport over man-made systems by generating the method of added mass efficiently [3].

The body motion traditionally adopted in previous robot fish can be represented by a travelling wave, applying a rigid mid-body and anterior, concentrating the undulatory motion to the posterior end of the lateral length, typically limited to <1/2 of the body length consisting of one positive phase and one negative phase [9][14]. Initiated at the center of mass, the posterior propagating wave smoothly increases in amplitude towards the tail. The observed common carp lateral excursion is 0.1 of the body length. The posterior confined kinematics of the carangiform is of the form [7]:

$$y_{body}(x,t) = \left(c_1 x + c_2 x^2\right)\sin(kx + \omega t) \tag{1}$$

where y_{body} is the transverse displacement of the body; x is the displacement along the main axis beginning at the nose; $k = 2\pi/\lambda$ is the wave number; λ is the body wave length; $\omega = 2\pi f$ is the body wave frequency; c_1 is the linear wave amplitude envelope and c_2 is the quadratic wave.

Free swimming robotic fish applying posterior confined displacements have shown significant kinematic parameter errors [9,10]. In particular, the lateral and thrust forces are not optimized around the center of mass, resulting in extensive anterior destabilization in the yaw plane, generated due to the posterior concentration of thrust. Consequently the inaccurate anterior midline parameters generate significant posterior kinematic displacement errors. Hence, the linear locomotive swimming motion over the full length of body has large matching errors in comparison to real fish leading to reduced propulsive force and a higher cost of transport. In addition the Sw may also be improved by applying large reaction forces (F_R) at the cost of energy consumption without certain gains in maximum velocity.

2.2 The Additional Swimming Motions

The full-body swimming pattern of the *iSplash* platforms proposed in [12] coordinates the anterior, mid-body and posterior body motions. This was based on intensive observation [12] and fluid-body flow interaction research [15,16] which lead to a greater understanding of the carangiform swimming motion. The coordinated full-body swimming motion was found to significantly reduce kinematic matching errors over the full body length and therefore attaining a low cost of transport. Our primary aim of this study is to further optimize this swimming motion by adjusting parameters, estimated to enhance performance.

In particular the displacements of full-body swimming pattern drive the anterior into the direction of recoil, reducing amplitude errors by optimizing the F_R of the propulsive elements. Furthermore the developed body motion enhances performance by producing a smooth transition of flow along the length of the body, effectively coordinating and propagating the anterior formed fluid flow interaction downstream. We can adjust the form in [9], an adaptation of (1) to generate the midline kinematic parameters of the full body displacements:

$$y_{body}(x,t) = \left(c_1 x + c_2 x^2\right)\sin\left(kx + \omega t\right) - c_1 x \sin\left(\omega t\right) \qquad (2)$$

where the values c_1, c_2, k, ω can be adjusted to achieve the desired posterior swimming pattern for an engineering reference.

The developed mechanical drive system can generate innumerable variations of the wave form, allowing a thorough investigation to identify the most efficient swimming motion within the limits of the structure. Some key deployed kinematic patterns are:

1. The Ostraciiform swimming mode [3], confining the kinematic displacements to the caudal fin.
2. The Anguilliform swimming mode [4]. Notably the fraction of the body length displaced of the full-body swimming motion of the *iSplash* platforms is equal to the anguilliform swimming mode (eel) but reflects changes in its kinematic form, as the full-body pattern applies an oscillatory motion to the anterior and mid-body segments and pivots the entire body around a single pivot point associated with the carangiform swimming mode [6].
3. Continued analysis of the full-body swimming motion by disregarding displacements of individual segments, to identify the contribution of each portion of the body length to the propulsion method.
4. Advancing and retarding the timing of individual segments within the sequences of the spatial and time dependent full-body wave motion.
5. Lastly, a series of tests investigating the effect of allowing individual segments to be passively moved by the surrounding fluid.

3 New Construction Method

3.1 Mechanical Design

The engineered platform 0.6m in body length is directly scaled from the structural link assembly of the first generation. This method provides an arrangement that has previously achieved high performance, in which we aim to improve further by precisely optimizing the kinematic parameters. Previously the accuracy of replicating the wave form parameters has been limited by hardware and material constraints. The structural approach of the multi-link servo assembly has typically been applied to optimize the kinematic displacements. This method is limited by force, frequencies, volume and mass distribution which are also typically confined to the posterior, reducing accuracy of the wave form [8],[9],[14]. Alternative methods deploying single continuous rotary actuators have measured large internal mechanical losses due to the complexity of the mechanism [7].

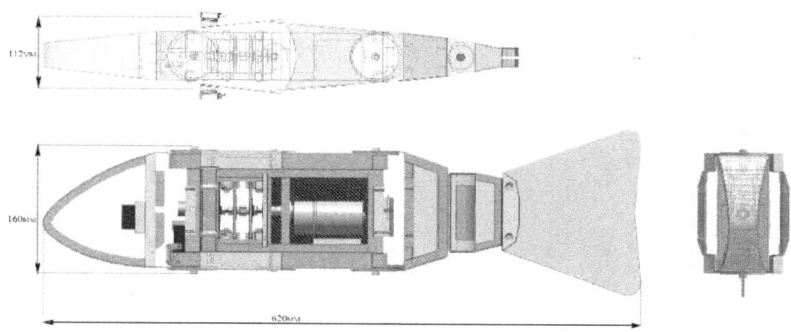

Fig. 2. 1-Plan; 2-Side; 3-Front view.

The presented prototype deploys an assembly with structural compliance combined with rigid discrete links. The arrangement distributes 3 degrees of freedom (DOF) and 1 passive DOF along the axial length to provide anterior, mid-body and posterior displacements and accurate midline curve alignment. The final posterior link V is coupled to a compliant scaled caudal fin, and is passively driven by four expandable tendons attached to the main chassis rear bulkhead, which can be adjusted experimentally to provide the targeted curves during free swimming at various frequencies, as achieved in [12]. Each of the discrete links across the body length can be configured to be actively displaced or held aligned with the centerline of the build. This development will allow analysis of each segments' contribution to the overall propulsion of the full-body wave form. In addition a significant aspect of drive system is the ability to attain free movement of individual links, allowing them to be passively moved by the surrounding fluid, as our subsequent work will investigate if the prototype is capable of extracting energy from the surrounding flow.

3.2 Power Transmission System

The power transmission system was developed to transfer power to and provide precise adjustments of the discrete links (with innumerable sequences) with the smallest mechanical loss as possible, whilst actuating at intensively high frequencies due to deploying a single continuous high torque actuator. The developed powertrain transmitting rotary power to linear oscillating sliders (with 13mm displacements) is illustrated in Figs. 3 and 4a. The three key sliders are directly driven by the three adjustable offset discs (which are secured to the drive shaft after adjustment), achieving equal power distribution, capable of transmitting power to discrete links across the full-body length from the lightweight tendons. Each of the discrete links was constructed with four adjusters increasing accuracy of the swimming patterns alignment by tuning the tendons (Fig. 4b). The developed mechanical drive system has high accuracy with unrestricted offset combinations, high structural strength and is small in size. This development is key to attaining an optimized swimming motion. The devised power transmission system required precision fitment of the chassis, crankshaft, cantilevers and linkages to avoid deadlock and reduce slide friction.

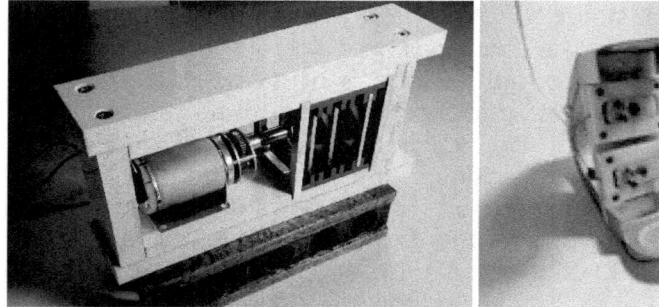

Fig. 4. Inner structure of the central drive system (a);   Adjusters for link I curve alignment (b).

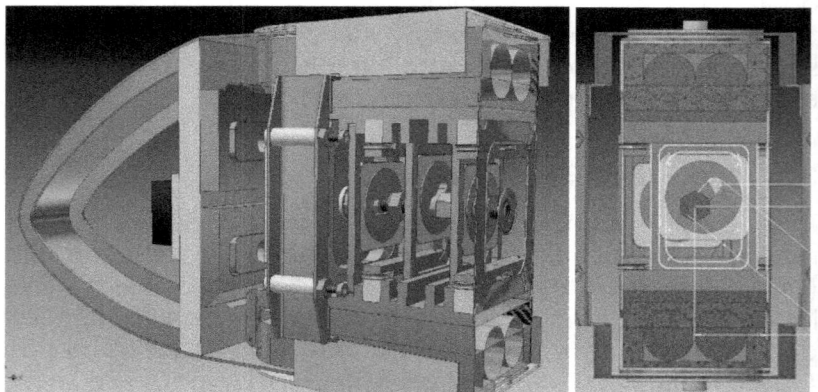

Fig. 3. Schematic drawing of the tail offset drive crank and linkage.

3.3 Fabrication

The modular prototype *iSplash*-OPTIMIZE shown in Fig. 5 was engineered as a morphological approximation of the common carp, fabricated using precision manufacturing techniques. The physical specifications are given in Table 1. A primary consideration of the development took into account the additional weight of the adjustable drive system, power supply and large electric motor (with dimensions of 85mm in length and 40mm in diameter), therefore no errors and excrescences in the geometry were required or kinematic parameters affected to compensate for the additional complexity of mechanical drive system. The geometric frame (the maximum cross section measured to be optimal at 0.2 of the body length) and midline camber was required to be accurate, as the outer profile of the coordinated full-body swimming pattern proposed in [12], represented by a deep camber aerofoil section (e.g. NACA (12)520) is estimated to aid the fluid flow interaction, producing greater locomotive speeds.

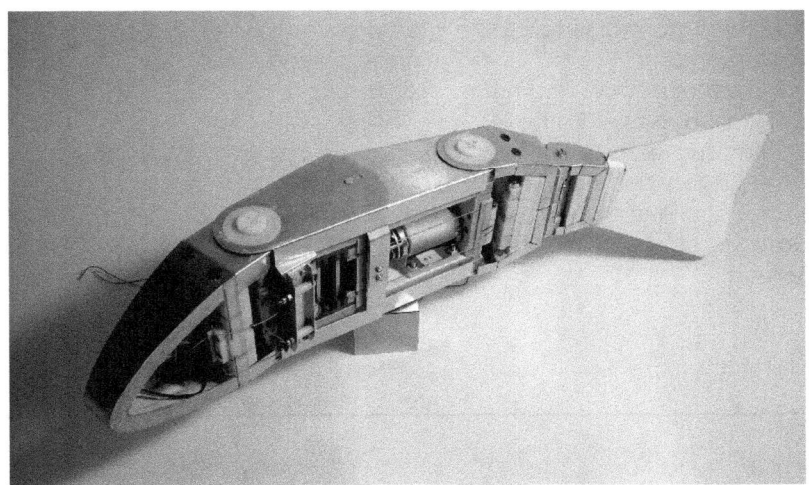

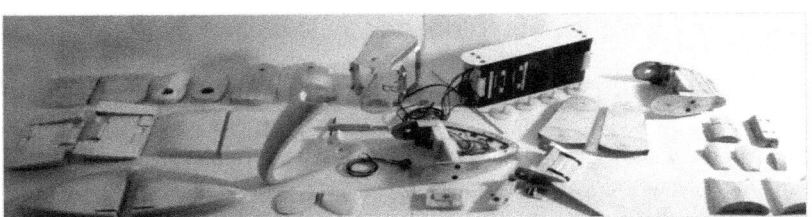

Fig. 5. *iSplash*-OPTIMIZE: Showing interchangeable parts of the modular build.

Table 1. Physical parameters of *iSpalsh*-OPTIMIZE

Parameters	Specific Value
Body size: m (L x W x H)	0.62 x 0.11 x 0.16
Body mass : Kg	0.9
Primary actuator:	Brushed DC motor
Power supply:	11.1v onboard LiPo battery
Manufacturing technique:	Precision engineering, machining
Primary swimming mode:	Linear locomotion
Additional maneuverability:	Yaw, pitch.
Additional control surfaces:	Pectoral fins
Tail Material:	Polypropylene
Thickness of caudal fin : mm	2.3
Caudal Fin Aspect Ratio: AR	1.6
Communication: (Additional)	2 x Zigbee 802.15.4, (27MHZ RF)
Tested signal distance: m	7.5
Microcontroller:	Arm Cortex M3 96Mhz
Data Storage:	SD card
Sensors:	Current, voltage, encoder, infrared, compass, accelerometer, gyroscope.
Materials:	Carbon Fiber, aluminum, stainless steel, acetal, low density foam.

The additional mass affected buoyancy, this was counteracted by improving the structures weight-to-strength ratio, deploying a combination of low density foam with carbon fiber layers, aluminum space frames and acetal inserts to strengthen sliding surfaces. This fabrication method achieved a structurally robust prototype, required for consistency of operation at high frequencies. In addition, natural buoyancy and open loop stability was achieved by positioning the large mass low and distributed across the main chassis (i.e. from DOF 2-3), therefore providing accurate density distribution properties to the first generation, as it is optimal for the principal pivot point of the carangiform swimming motion to be positioned within the range of 0.15-3 of the body length.

Additional mobility mechanisms were devised for autonomous locomotion: (i) to turn within the vertical plane. Two rigid morphological approximations of pectoral fins were developed and positioned at the leading bulk head of the main chassis, actuated by a single servo motor (with 180° actuation); (ii) to turn within the horizontal plane. A mechanism was developed to adjust the swimming pattern to generate the form of a C-sharp turn [4], by offsetting the posterior tendons using adjustable cross sliders actuated by a single servo motor.

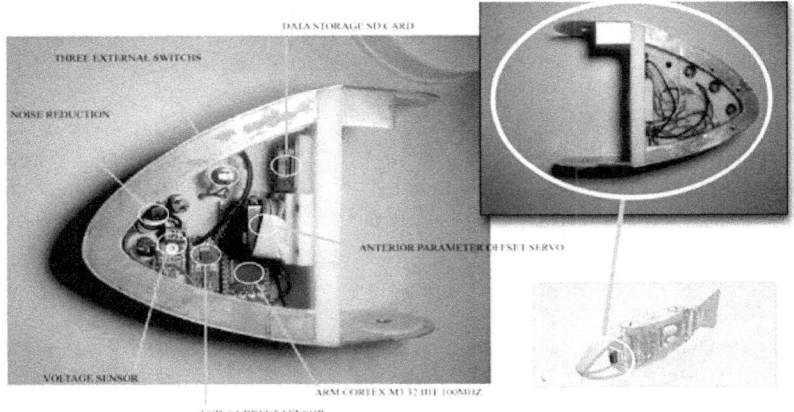

Fig. 6. Inner Structure of link I showing the installed electrical system.

4   Embedded System

A modular electrical system was built, employed to measure and control the robotic fish motion, positioned within link I shown in Fig. 6. The first study aims to measure various kinematic sequences in an attempt to reduce the cost of transport during linear locomotion. The central processor is a 32bit 96MHZ ARM Cortex-M3, which samples and filters data from multiple deployed sensors (i.e. incremental encoder, current, voltage and an inertial measurement unit (IMU)), is employed to measure the energy economy and the extent of destabilization at various frequencies and swimming patterns. The energy consumption of *iSplash*-I was sampled by external electronics. The new onboard electronic system and free swimming robotic fish allows for measurements unaffected by a tethered power supplies or support beams. The data can be logged, onboard by the SD module or sent instantaneously (tested up to a distance of approximately 7.5m) to an external PC for analysis by two onboard 802.15.4 wireless radios.

Our planned future work is to achieve autonomy at high locomotive speeds. The system has been installed with four infrared sensors (tested to produce accurate data up to approximately 1m) and two control surface actuators so that the central processor will process data of the state of orientation, surrounding environment, energy economy and frequencies rate, and therefore perform decision making and produce the required control signals. The structure is shown in Fig. 7. This will be great challenge as locomotive speeds increase due to the deployed limited sensors.

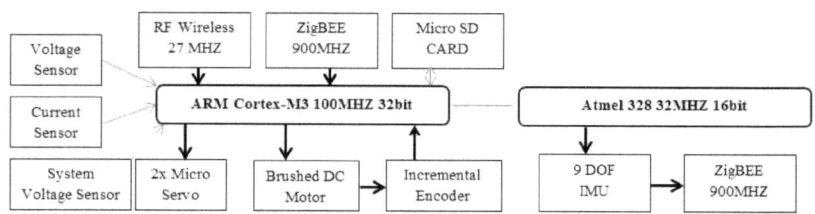

Fig. 7.   The structure of the energy and stability measurement and actuator control system;

5 Experimental Procedure and Results

5.1 Mechanical Energy Transfer

A series of experiments were undertaken in order to show the feasibility of the new prototype, evaluating performance in terms of kinematic parameters, robustness, open loop stability and energy consumption at frequencies within the range of 5-20Hz. The test results are given in Table 2. The embedded system was employed to sample data reaffirmed by external electronics. The measurements were required to be averaged over many cycles to increase the accuracy of data, once consistency of operation was achieved. The initial experimental testing highlighted areas of required development. During actuation at high frequencies we found that the chosen materials and configuration enduring high forces failed as the build design was not suitable for the large increase in scale, which produced a significant increase in pressure in comparison to the first build. Rebuilds were undertaken and the appropriate materials suitable for intensively high frequencies within an aquatic environment were found. The rebuilds took into consideration mass and volume distribution and material properties, so that open loop stability was realized, which is a requirement for precise optimization of the swimming motion during linear locomotive. For the development of autonomy with mobility in yaw and pitch it is required that the weight is greatly concentrated at the center mass [17].

Once consistency of operation was achieved the devised mechanical drive system was found to be very robust, showing no signs of structural failure throughout experimental testing, whilst actuating at intensively high frequencies for long periods in air at 10Hz and short test periods of actuation at 20Hz in water.

The power supply will contribute to a significant portion of the total mass therefore an efficient energy transfer is required for high performance. In Table 2 the average energy economy in relationship to driven frequency is shown, during non-productive power consumption (i.e. while operating in air). This comparison was used to measure the value of the increased resistance during actuation of each discrete link and during various link sequences. We measured the energy economy over many cycles, as the inertia of the oscillatory motion produces fluctuating readings within a cycle. Operation whilst actuating all links at 10Hz at their maximum lateral excursion resulted in a small increase in energy consumption to 9.7W compared to the unloaded motor at 6.1W and the mechanical drive adjusters at 6.4W. Actuation of link I measured an energy consumption of 7.2W, links II and III measured 6.4W, link IV and V measured 6.8W. We can see that the developed mechanical drive system transfers continuous rotary motion to discrete oscillatory links across the platform with little internal mechanical loss. Therefore we can calculate that the prototype may be capable of carrying an onboard power supply within the current geometric frame.

Table 2. Experimental Test Results of *iSpalsh*-OPTIMIZE

Parameters	Specific Value
Max Frequency tested in air: Hz	10
Max Power Consumption Motor No Load: W	6.1
Max Power Consumption Mechanical Drive Adjusters: W	6.4
Max Power Consumption only link I: W	7.2
Max Power Consumption only link II and III: W	6.4
Max Power Consumption only link IV and V: W	6.8
Max Power Consumption all links: W	9.7
Link I – Max Head displacement: m (Joint angle (JA)°)	0.060 (22)
Link II – Max Mid-body displacement: m (JA°)	0.020 (14)
Link III – Max posterior link displacement: m (JA°)	0.020 (11)
Link IV – Max Tail displacement: m (JA °)	0.048 (28)
Link V – Max Tail displacement: m (JA°)	0.040 (41)
Total anterior amplitude: m (of the body length°)	0.060 (0.1)
Total posterior amplitude: m (of the body length°)	0.167 (0.3)

5.2 Kinematic Parameters

The midline kinematics of the full-body swimming pattern were tracked at 50 frames per second during actuation in air to provide the amplitude values of the anterior, mid-body and posterior, for comparison with real fish and *iSplash*-I. Good agreement with live fish kinematic data is a difficult task and current free swimming robotic fish have shown excessive head and tail amplitude errors during locomotion. All links were tested at half the maximum frequency (i.e. 10Hz) so that the build was not damaged due inertia forces. Link I was able to attain a maximum amplitude of 0.1 (0.06m) of the body length, measured from the midline to the maximum lateral excursion at a turning angle of 8°. Link II and III attained a maximum amplitude of 0.03 (0.02m) with 14° and 11° respectively. Link IV attained a maximum amplitude of 0.08 (0.048m) with 28° and Link V attained a maximum amplitude of 0.07 (0.040m) with 41°. The maximum lateral head (i.e. 0.1) and tail (i.e. 0.27) excursions generated are significantly greater than the observed common carp, and *iSplash*-I (Notably the tail amplitude of *iSplash*-I was measured to increase performance with larger values than the common carp at 0.1 of the body length able to attain values of 0.17 (0.044m) due to achieving anterior stabilization). We can assume that applying the maximum attained amplitudes of *iSplash*-OPTIMIZE during locomotion will generate negative propulsive forces, therefore providing adequate displacements to find the optimized swimming pattern. Significantly, we were able to adjust the mechanical drive system to generate numerous link sequences (with innumerable combinations of the cross sliders), therefore producing accurate swimming patterns during non-productive actuation.

6 Conclusion and Future Work

This paper details the design, fabrication and mechanical efficiency tests for a bio-robotic marine vehicle, showing its feasibility as a platform to accurately optimize the carangiform straight line swimming motion over previous methods. Devised to reduce the kinematic errors by precisely tuning the reaction forces of the propulsion elements during locomotion, the developed mechanical drive system has shown the capability to generate accurate spatial and time dependent discrete link sequences during non-productive actuation at 10Hz with a small energy consumption of 9.7W over a non-loaded motor at 6.1W. The swimming patterns at high frequencies were attained by realizing a powertrain with high accuracy unrestricted disc offset combinations, small in size and with high structural strength, able to transfer power to and provide precise adjustments of the oscillatory discrete links across the full assembly from a single continuous rotary actuator. The details of the onboard electrical system were given showing its practicality for measuring and controlling the energy consumption, stability and mobility by deploying control mechanisms for the actuation of pectoral fins and adjustment of the linear swimming motion to generate the form of a C-sharp turn.

The achievements of conducted mechanical tests have indicated the following significant aspects to improve the next generation: (i) The devised adjustable mechanical drive system is suitable for a reduction in scale, relating to reduce forward resistance; (ii) An autonomous parameter adjustment system may be deployed, utilizing a series of servo motors, to attain swimming motion adjustment across the range of frequencies during locomotion; (iii) The robust structure may allow a continued raise in frequencies; (iv) The compact drive mechanism is suitable to distribute power to additional links, increasing redundancy without greatly increasing mechanical complexity and the geometric frame.

Our future research will now focus on the experimental testing of *iSplash*-OPTIMIZE, a series of experiments will be conducted in order to verify the prototype by evaluating the locomotive performance in terms of kinematic parameters during linear locomotion, speed, force, energy consumption, horizontal and vertical plane mobility and autonomous operation at intensively high frequencies within the range of 5-20Hz.

Acknowledgments

Our thanks go to Richard Clapham senior for his constant technical assistance towards the project.

References

1. P. R. Bandyopadhyay, "Maneuvering hydrodynamics of fish and small underwater vehicles," Integr. Comparative Biol., vol. 42, no. 1, pp. 102–17, 2002.
2. J. Lighthill, "Mathematical Biofluiddynamics", Society for Industrial and Applied Mathematics, Philadelphia, 1975.
3. J. J, Videler, "Fish Swimming", Chapman and Hall, London, 1993.
4. J. Gray, "Studies in Animal Locomotion," J Exp Biol 10, 88-104, January 1933.
5. G. S. Triantafyllou, M. S. Triantafyllou, and M. A. Grosenbauch, "Optimal thrust development in oscillating foils with application to fish propulsion," J. Fluids Struct., vol. 7, pp. 205–224, 1993.
6. M. Nagai. "Thinking Fluid Dynamics with Dolphins," Ohmsha, LTD, Japan, 1999.
7. D. S. Barrett, M. S. Triantafyllou, D. K. P. Yue, M. A. Grosenbaugh, and M. J. Wolfgang, "Drag reduction in fish-like locomotion," J. Fluid Mech., vol. 392, pp. 183–212, 1999.
8. J. Yu, M. Tan, S. Wang, E. Chen. "Development of a biomimetic robotic fish and its control algorithm," IEEE Trans. Syst., Man Cybern. B, Cybern., 2004,34(4): 1798-1810
9. J. Liu and H. Hu, "Biological Inspiration: From Carangiform fish to multi-Joint robotic fish," Journal of Bionic Engineering, vol. 7, pp. 35–48, 2010.
10. P. Valdivia y Alvarado, and K. Youcef-Toumi, "Modeling and design methodology for an efficient underwater propulsion system", Proc. IASTED International conference on Robotics and Applications, Salzburg 2003.
11. R. Bainbridge, "The Speed of Swimming of Fish As Related To Size And To The Frequency And Amplitude Of The Tail Beat", J Exp Biol, 35:109–133, 1957.
12. R.J. Clapham and H. Hu, "*iSplash*-I: High Performance Swimming Motion of a Carangiform Robotic Fish with Full-Body Coordination," IEEE International Conference on Robotics and Automation, May 31 - June 7, 2014, Hong Kong, China.
13. P. W. Webb, "Form and function in fish swimming," *Sci. Amer.*, vol. 251,pp. 58–68, 1984.
14. P. Nilas, N. Suwanchit, and R. Lumpuprakarn, "Prototypical Robotic Fish with Swimming Locomotive Configuration in Fluid Environment," Proceeding of the International Multi-Conference of Engineers and Computer Scientists 2011, page 15 -17, March 16-18, 2011.
15. M. W. Rosen, "Water flow about a swimming fish," China Lake, CA, US Naval Ordnance Test Station TP 2298, p. 96, 1959.

16. M.J. Wolfgang, J.M. Anderson, M.A. Grosenbaugh, D.K. Yue and M.S. Triantafyllou, "Near-body flow dynamics in swimming fish," September 1, 1999, J Exp.
17. G.V. Lauder and E.G. Drucker, "Morphology and Experimental Hydrodynamics of Fish Control Surfaces," IEEE J. Oceanic Eng., Vol. 29, Pp. 556–571, July 2004.

iSplash Robotics

Published by *iSplash* Robotics UK
www.isplash-robotics.com
Copyright © 2016

Published by *iSplash* Robotics 2016
Illustration by Richard James Clapham PhD

All rights reserved. No part of this publication may be reproduced, stored in a retrieval system or transmitted in any form or by any means, electronic, mechanical, photocopying or otherwise without the prior permission of Natural Classics.

ISBN:

Thank you for your purchase.

www.ingramcontent.com/pod-product-compliance
Lightning Source LLC
Chambersburg PA
CBHW070342190526
45169CB00005B/2008